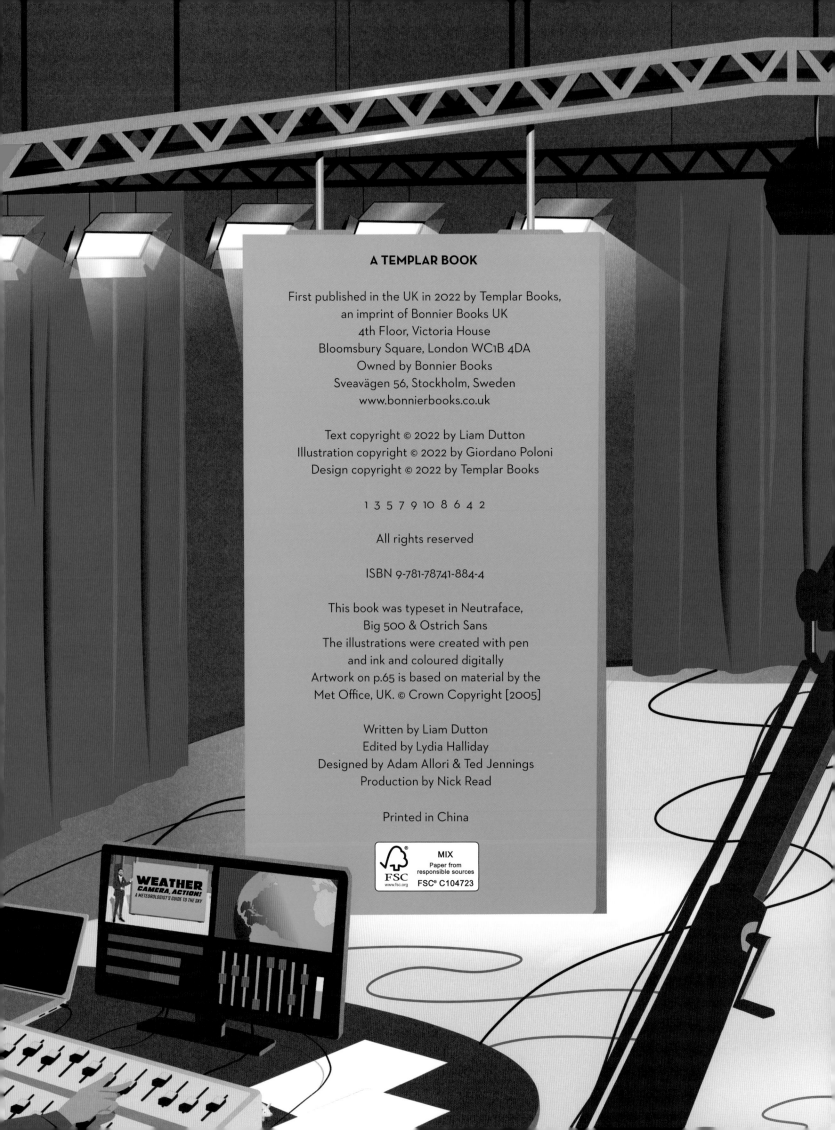

A TEMPLAR BOOK

First published in the UK in 2022 by Templar Books,
an imprint of Bonnier Books UK
4th Floor, Victoria House
Bloomsbury Square, London WC1B 4DA
Owned by Bonnier Books
Sveavägen 56, Stockholm, Sweden
www.bonnierbooks.co.uk

1 3 5 7 9 10 8 6 4 2

ISBN 9-781-78741-884-4

This book was typeset in Neutraface,
Big 500 & Ostrich Sans
The illustrations were created with pen
and ink and coloured digitally
Artwork on p.65 is based on material by the
Met Office, UK. © Crown Copyright [2005]

Written by Liam Dutton
Edited by Lydia Halliday
Designed by Adam Allori & Ted Jennings
Production by Nick Read

Printed in China

WEATHER
CAMERA, ACTION!
A METEOROLOGIST'S GUIDE TO THE SKY

Written by
Liam Dutton

Illustrated by
Giordano Poloni

templar
books

MEET THE AUTHOR

Ever since I was around six years old, I've always been interested in the weather. I don't remember exactly how it happened, it just did. Whether it was peering out of the window looking for snow or being in awe of flashes of lightning flickering across the sky, weather has always fascinated me. All of these amazing things going on around us that we have no control over, yet they affect our lives in so many ways.

As I grew up, this fascination never went away. If anything, it grew stronger. By the time I was a teenager, I knew that I wanted weather not just to be a personal interest, but also a career. So, after focusing on science, geography and maths-related subjects through college and university, I fulfilled my ambition of becoming a qualified **meteorologist** – someone who studies and predicts the weather. I also love talking to people about it, and so I became a weather presenter on TV.

Since 2003, as well as becoming a Met Office-trained meteorologist, I have presented the weather on both the BBC and Channel 4 television networks. Over the years, I've broadcast to hundreds of millions of people around the world – on TV, radio and online. I also went viral on the internet for saying a very, very, very long place name during one of my weather forecasts!

As well as presenting the weather, I am passionate about explaining the science behind it and what it means. The need to understand what is going on around us has never been more important – especially at times of **extreme** weather and as Earth experiences the impacts of **climate change**. That's why I have written this book.

So, welcome, and let our journey through the wonderful world of weather begin!

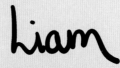

JET STREAMS

AN ATMOSPHERIC SUPERHIGHWAY

Imagine looking down on Earth from the deep darkness of space: its vast blue oceans, golden deserts, luscious forests, all topped and tailed by crisp, ice-covered poles.

But there's more. Clouds, and lots of them – big and small, swirls and blobs, bright and white – whizzing around on an atmospheric superhighway known as the jet stream.

Let's Get Moving!

Jet streams not only determine how and where weather systems form, but they also shift them around. For example, an area of cloud, rain and wind may start in eastern Canada, but could end up thousands of miles away in Europe – changing as it moves.

Swirling from west to east as fast as 400 kph, jet streams are powerful ribbons of wind found at the same **altitude** as aeroplanes.

There are two main types of jet stream: the **polar** jet streams and the **subtropical** jet streams. The northern and *southern* halves of Earth both have one of each.

Hot Pursuit

Temperature differences in the **atmosphere** cause air to rise, fall and move around our planet, and it's the contrast between cold air at the poles and warm air at the **equator** that leads to jet streams.

The bigger the contrast, the faster the jet stream. Therefore, it moves fastest in autumn and winter, when the contrast between the increasingly cold poles and constantly warm equator gets bigger. In summer, it moves much slower, as the temperature contrast decreases.

*Polar jet streams circle at **mid-latitudes**. They have a big influence on our weather, so you might hear about them in forecasts.*

Subtropical jet streams are more central, near the tropics. They usually have less influence on our weather, but when they join forces with polar jet streams, things can get lively!

EQUATOR

Going My Way?

Planes can fly faster by hitching a ride on a jet stream. If they're moving in the same direction, the jet stream helps to carry the plane along and gives it a speed boost.

Don't Slow Down

Normally, jet streams smoothly flow around the Earth, keeping weather changeable. But when they bend too much, weather systems can slow down or even get stuck, causing spells of rain, snow, heat or cold that last for weeks or even months. This is how extreme weather can happen...

WIND

STIRRING THE AIR

Although we can't actually see the wind, we can see the big effect it has around us. But where does it all come from, and how does it move the weather?

Under Pressure

Wind is caused by differences in **air pressure**: the weight of air molecules pressing down on the ground. Imagine you're climbing a ladder: this is usually pretty easy, because nothing's pushing you down – that's low pressure. But if you tried to climb with a full backpack on, it would be much harder because it's pushing you down, and that's high pressure.

Areas of **high pressure** generally bring calm, settled weather. Air sinks towards Earth's surface, spreading outwards when it reaches the ground.

Areas of **low pressure** bring unsettled weather, and air rises upwards – away from Earth's surface.

Wind always blows from high to low pressure and it's this movement of air between them close to the surface that creates wind.

I often mention high and low pressure in my weather forecasts. It's a handy way to give everyone a clue as to whether they'll have to hold on to their hats or not!

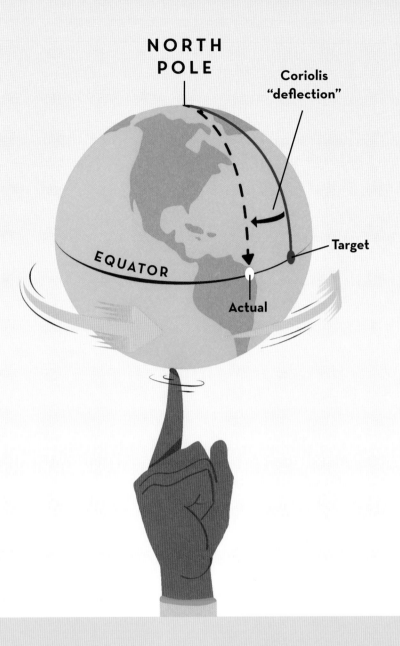

NORTH POLE

Coriolis "deflection"

Target

EQUATOR

Actual

Wherever the Wind Blows

If you know the direction that the wind is coming from, you can work out the weather that it's bringing your way. For example, if you live in Europe and have a northerly wind in winter, it'll come from the **Arctic**, so it'll bring cold air and possibly snow. If there's a southerly wind in summer then it'll come from Africa, bringing hot air, clear skies and perhaps even thunderstorms.

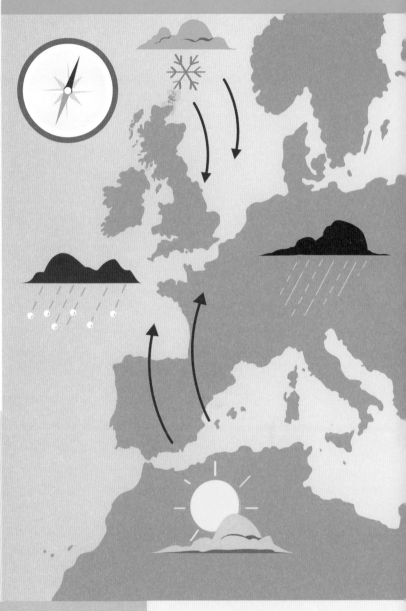

You Just Spin Me Around

In the **northern hemisphere**, wind blows clockwise around high pressure and anticlockwise around low pressure. But in the **southern hemisphere**, that's reversed! This is due to the **Coriolis effect**, where the spinning motion of Earth causes wind to be deflected (curve) to the right in the northern hemisphere and to the left in the southern hemisphere.

How Are Your Grades?

The strength of the wind depends on how much the air pressure changes over distance. This change is called the 'pressure gradient'.

If the air pressure changes a lot over a short distance, then you're in for strong wind.

If the same change happens over a longer distance, the wind doesn't move so fast, so it's weaker.

AIR MASSES

WEATHER ON THE MOVE

Every time we step outside, the air determines how we feel and the clothes we wear: warm or cold, wet or dry, season to season and day to day. So why does it change? That's up to air masses.

Weather on the Move

Air masses are huge bodies of air that roam our planet, covering whole countries or even continents and reaching high into the atmosphere. Each air mass has a particular temperature and humidity (moisture content) which determines how it's described and the weather it brings.

How to Name the Air

To make them easy to recognise, we describe air masses by which **climate zone** they come from (polar, Arctic, **tropical**, **equatorial**), and if they start over land (**continental**) or sea (**maritime**). An **air mass** from the Caribbean Sea in the tropics would be warm and moist, so that's 'tropical maritime'. An air mass from Canada would be cold and dry, so 'polar continental'.

Making an Air Mass

Even rooms in your house can have different air masses. After a shower, your bathroom's air would be warm and **humid**. However, when you go to an air-conditioned bedroom afterwards, it would be cool and dry. If you put a fan on, you could move the cool, dry air into the warm, humid room and change how it feels. Air masses move and change too. A polar continental air mass moving over the sea can pick up moisture and turn polar maritime. But when air masses get stuck where they wouldn't normally be, they can really stir up trouble...

War of the Winds

In the winter of 1962–63, the United Kingdom had its coldest winter in over 200 years, as Arctic air lingered for two months. I remember my mum telling me about it. She said that it was a winter that never seemed to end! Lakes and rivers iced over, and each time tropical maritime air from the Atlantic Ocean tried to move in, the collision with the Arctic air led to huge snowfalls. A **blizzard** built snowdrifts up to 6 metres high, cutting off villages and freezing farm animals to death. The battle raged for much of the winter before the Arctic air finally surrendered and eased away.

TROPICAL WEATHER

DODGING THE DOWNPOURS

Vast forests swelter in the tropical regions of our planet, their hot air full of moisture. This moisture helps the rainforests grow, but also builds towering clouds and thunderstorms, whose downpours drench the swamps and rivers to keep them full for animals and vegetation. This weather isn't driven by jet streams but is caused by the intertropical convergence zone (ITCZ).

Cutting the World in Half

The **ITCZ** is a zone of low pressure around Earth. If you look at a **satellite** photo, you can see it in the bright white clouds circling the middle. Each year, the ITCZ drifts north and south between the **Tropic of Capricorn** and the **Tropic of Cancer**, across a line called the equator which divides the planet into northern and southern halves, known as hemispheres. The Tropic of Cancer and Tropic of Capricorn mark the edges of Earth's tropical climate zone, and are 2,600 kilometres north and south of the equator.

Who Needs Four Seasons?

In the tropics, there are two clear seasons: a wet **monsoon** season, and a dry season. These can last for weeks or months. Wet seasons arrive late spring into summer, because the ITCZ follows the sun's heat. In the northern hemisphere, this is from May to July, and in the southern hemisphere, it's from November to January.

Clue's in the Name

But what causes the ITCZ? Surface winds from each hemisphere crash together – or converge – forcing air up into the sky. The air here is hot and humid from the heat of the sun and warm oceans, but as it rises, it cools and condenses into water droplets (like hot steam on cold glass) or even freezes into ice crystals. These droplets and crystals are the building blocks for huge clouds, which bring showers, heavy rain and thunderstorms.

The Calm Before...

The ITCZ weather might sound calm and simple, but as it moves north and south, its monsoon rains can build the thunderstorms that birth powerful, dangerous weather events like **hurricanes**, typhoons and tropical cyclones...

Stuck in the Doldrums

Weather can sometimes be calm in the ITCZ with very little wind. Historically, sailors called this area 'the **doldrums**', and the light winds were a nightmare when they needed strong **gusts** to blow them along. That's where we get the phrase 'stuck in the doldrums'.

SOUTH ASIA MONSOON

GETTING DOWN WITH THE DELUGE

Like the flick of a switch, South Asia can go from blazing sun and stifling heat to rains that pour for months. Dry, dusty roads are quickly soaked, turning streets into rivers; brown, barren landscapes burst to green with emerging vegetation. That's the power of monsoon rains.

But How Soon?

A monsoon is a seasonal wind that brings the same type of weather for a prolonged period. The most well known one is the South Asia monsoon. Usually arriving across Thailand, Myanmar and Sri Lanka at the end of May, it's a wet season that then takes around six weeks to spread north across India, Bangladesh and Pakistan.

Rainy Day Fund

The monsoon is a vital water supply, providing parts of India with 80 per cent of their annual rainfall. Farmers rely on the monsoon for crops to sell internationally – such as rice, tea and cotton – so if the rain arrives late or underdelivers, it can cause a shortage that drives up prices around the world.

Bringing the Heat

The South Asia monsoon is caused by the ITCZ (see p.14–15) drifting north during summer. Before it arrives, months of clear skies and intensifying sunshine build up heat over the land. Temperatures widely reach 35-45 degrees Celsius, and dangerous **heat waves** can cause **heatstroke**, especially in cities. The air rises, creating an area of low pressure, and a southwesterly wind blows in from the Indian Ocean and fills the region with a vast, warm, humid air mass. It feeds the monsoon with moisture to create the pouring rains.

The Harder They Fall

Monsoon rains are greatest over mountainous areas, where the moisture-laden air is forced higher, making the rainfall heavier than in low-lying areas. Unsurprisingly, this means the wettest places on Earth are in India, with some locations having nearly 12 metres of rain each year: twice the height of a giraffe!

Breaking the Spells

Even though monsoon rains come in short spells, rather than falling non-stop, they're visible even from space as a bright band of clouds. Eventually, the monsoon rains drift south from Pakistan and northern India and leave South Asia by November. The dry season begins, and the cycle is complete.

Give and Take

Even though the monsoon rains are crucial for supporting life, they regularly cause deadly flooding and **landslides**. Steep slopes of soil become so **waterlogged** that they collapse into destructive torrents of mud that can sweep away trees and houses.

ATLANTIC HURRICANES

Hurricanes can be catastrophic, with winds roaring like a jet engine and powerful enough to blow down buildings. Storm surges – sea water pushed inland by hurricane-force winds – can swell to flood whole cities. Skies turn dark and menacing: something unstoppable is coming.

Hurricane Season

In the Atlantic Ocean, hurricane season runs from 1 June to 30 November, with the busiest time for hurricanes typically being mid-September. The number of hurricanes each year varies greatly. Some years can have as many as 10! Life for people in these regions carries on as normal, but they are always alert and prepared.

PLEASE DRIVE CAREFULLY!

Stirring Up A Storm

ROUTE 75

A hurricane is born from a 'tropical disturbance': a cluster of showers and thunderstorms around an area of low pressure over the ocean. In order to grow into a hurricane, they need:

Sea surface temperatures of 26.5 degrees Celsius or above to provide energy and rising moisture.

Winds at the surface need to converge and collide, so air is forced up to form thunderstorm clouds.

Winds that have a similar speed and direction with increasing altitude, otherwise the storm clouds will be ripped apart.

A little distance away from the equator, so the spinning effect of Earth can get the storm rotating.

If all of these factors come together, then the winds will start to spin the cluster. If they are 62 kilometres per hour or less, it's called a **tropical depression**; at 63-118 kilometres per hour, it's a **tropical storm**. But when winds hit 119 kilometres per hour (74 mph) or more, you've got a hurricane.

Out With a Bang or a Whimper?

Hurricanes can travel thousands of kilometres and last for weeks, but some only survive a few days. As soon as they lose their warm-water energy source – either by making landfall or moving over colder water – they weaken and die. Hurricanes tend to fade gradually, but if they collide with mountains or a **weather front**, they can get ripped apart and meet their end swiftly.

Land-Ho!

Once a hurricane has formed, it moves westwards on the trade wind – a wind that blows just north of the equator – towards the Caribbean, Central America or the United States. There it can hit land, an event known as '**making landfall**'.

Not all hurricanes make landfall. Some stay over the ocean. These are sometimes called '**fish storms**' because they bother no one apart from the fish.

Test Your Strength

A hurricane's strength is measured by the Saffir-Simpson scale, which rates it on a scale of one to five – with five being the strongest. During their life cycle, these hurricanes can expend as much energy as 10,000 nuclear bombs!

HURRICANES, TYPHOONS AND TROPICAL CYCLONES

SAME BUT DIFFERENT

Hurricanes, typhoons and tropical cyclones form in the same way, look the same and create the same dangerous weather. So why do they have different names? This is one of the most common questions that I'm asked when these storms make the headlines. The answer lies in which of our planet's tropical oceans they form.

STORMS THAT FORM IN THE ATLANTIC AND EASTERN NORTH PACIFIC OCEANS ARE CALLED HURRICANES.

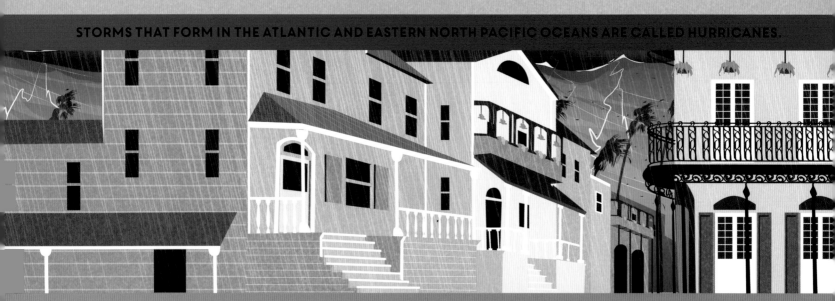

IF STORMS FORM IN THE WESTERN NORTH PACIFIC OCEAN, THEY ARE CALLED TYPHOONS.

IF STORMS FORM IN THE INDIAN OCEAN AND AROUND AUSTRALASIA, THEY ARE CALLED TROPICAL CYCLONES.

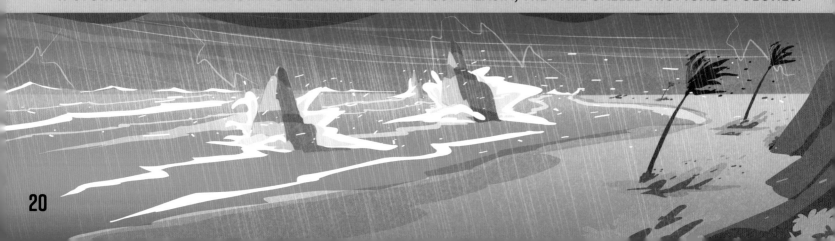

Eye of the Storm

From space, each hurricane, typhoon or tropical cyclone is a big swirl of cloud, typically 480-640 kilometres wide, with a clear circular zone in the middle. In this cloudless 'eye of the storm', the weather is calm with little wind or rain. There's even a clear sky. However, the cloud immediately surrounding this area – the 'eye wall' – is where you'll find the most powerful and damaging winds.

Remember the Coriolis effect? Hurricanes, typhoons and tropical cyclones follow those rules too, rotating in different directions in each hemisphere: anticlockwise in the northern and clockwise in the southern (see page 11).

Keep to the Side

The further out from the centre of the storm you go, the weaker the winds become. This means that not everywhere the storm hits will experience the most damaging winds.

What Did You Call Me?

In order for these storms to be easily identified and described – and to warn people that they are coming – they are given names. But who names the storms? Well, the **World Meteorological Organization** has a committee where weather forecasters from each storm-affected area meet. They all agree on around five lists of different names for each respective region.

These lists often work through the alphabet, alternating between male and female names. The lists are usually reused over time. However, when a storm has been extremely damaging, with significant loss of life, the name is removed from the list and replaced.

21

The New Orleans Bulletin

VOL. 18   NO. 452   TUESDAY, AUGUST 30, 2005   80 PAGES   50 CENTS

HURRICANE KATRINA

DEADLY FLOODING HITS NEW ORLEANS

Hurricane Katrina hit the southern US in August 2005. It was one of the deadliest hurricanes the country has ever seen, killing around 1,800 people and forcing over a million people to leave their homes.

Katrina started off as a tropical depression – an area of thunderstorms and gusty winds – south of the Bahamas on 23 August. It then became a hurricane on 25 August, before first moving over land north of Miami, Florida. In the following days, Katrina moved over the warm waters of the Gulf of Mexico. This gave the storm enough energy to reach category five on the Saffir-Simpson scale (that measures a hurricane's strength), with winds of 282 kilometres per hour.

On 29 August, Katrina made its final devastating landfall along the coast of Louisiana as a category three hurricane, with sustained winds of 200 kilometres per hour.

Hurricane Katrina sent a **storm surge** of ocean water kilometres inland, flooding the states of Mississippi and Louisiana – including the city of New Orleans. Impacts from the storm surge were made worse because the **levees** (man-made structures that protect low-lying areas from floods) failed. The region was swamped with flood water.

The damage and loss of life caused by the hurricane was so extreme that 'Katrina' has been retired from the list of names given to storms, replaced by 'Katia'.

KATRINA FACTS

⚡ *Most costly hurricane on record at US $108 billion, in terms of repairing and rebuilding everything that was damaged.*

⚡ *1.3 million people **evacuated** from southeast Louisiana, including 400,000 from New Orleans itself.*

⚡ *80 per cent of New Orleans was flooded, with some areas 4.6 metres underwater – the height of a double-decker bus.*

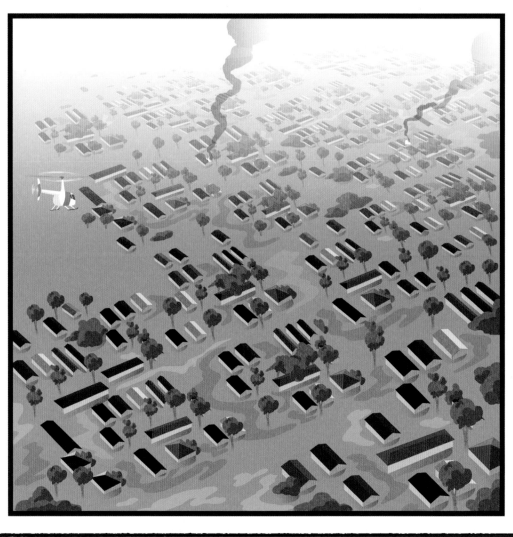

Super Typhoon Haiyan

The Philippines is no stranger to tropical storms, with around 20 affecting the region each year, mostly from July to October. Super Typhoon Haiyan, known locally in the Philippines as Yolanda, made landfall on eastern Samar Island on 8 November 2013.

It was one of the most powerful storms on record, with sustained winds of 314 kilometres per hour and even stronger gusts. This is equivalent to a category five hurricane on the Saffir-Simpson scale. Winds this powerful destroy almost everything in their path, leaving the affected area uninhabitable for weeks or months.

Haiyan ploughed through the Philippines, killing approximately 6,300 people and affecting around 14 million. It flattened buildings and left many homeless. Hardest hit was Tacloban, a city on the island of Leyte, where local estimates suggested that 90 per cent of the city was destroyed.

Like Hurricane Katrina, a major cause of damage was a storm surge which washed away homes and piled up cars in the streets. **Torrential rain** only made this worse, with 280 millimetres of rain falling in just 12 hours, triggering deadly landslides and **flash floods**.

03

00:44:23

Super Typhoon Haiyan facts

- *There was an estimated US $5.8 billion of damage caused.*
- *Around 1.1 million homes were damaged or destroyed.*
- *It's the deadliest typhoon on record for the Philippines.*
- *Huge areas of crop-producing land were destroyed, affecting food production.*

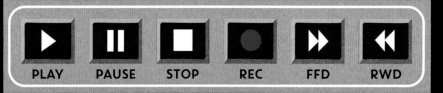

PLAY   PAUSE   STOP   REC   FFD   RWD

STRATUS, CUMULUS, CIRRUS

Cirrus

Cirrocumulus

Cumulonimbus

Altocumulus

Cumulus

A CLOUD SPOTTER'S GUIDE

Take a look out of your window. The chances are, there's at least a few clouds drifting by, and their size, shape, movement and colour set the mood for us down below. Fluffy, white cumulus clouds lift our spirits and entice us outside. Big, dark cumulonimbus clouds threaten rain – scaring us back indoors.

Floating by...

Clouds are massive collections of water droplets and ice crystals that weigh so little that they float in the air. Almost all clouds are found in the lowest layer of Earth's atmosphere – the **troposphere** – because this is where nearly all of the moisture is located. Low clouds contain mostly water droplets, clouds higher up have a mixture of water droplets and ice crystals, and the highest clouds are made entirely of ice crystals.

The Cloud Atlas

If you're having a hard time keeping track of the different types of clouds I've listed here, you're not alone! In 1802, a pharmacist from London named Luke Howard thought exactly the same thing. As an amateur meteorologist, this inspired him to come up with a system for naming the various cloud types, in order to have a common system that everyone could understand. He proposed classifying clouds into three main categories, based on their appearance.

Stratus/strato – a flat or layered smooth cloud.

Cumulus/cumulo – a heaped or puffy cloud.

Cirrus/cirro – a high-up wispy cloud.

Luke's proposal was so popular that it was accepted by meteorologists around the world. This led to the first *International Cloud Atlas* being published in 1896 – a book which is still produced and updated to this day.

Understanding Cloud Names

The World Meteorological Organization extended Luke Howard's cloud classifications to make 10 main cloud groups, called *genera*.

They did this by adding the descriptions 'alto', meaning 'medium level', and 'nimbus/nimbo', meaning 'rain-bearing'. By combining these terms, we can describe and understand each cloud, for instance:

Cumulo + nimbus = cumulonimbus
(a heaped, puffy and rain-bearing cloud).

Cirro + stratus = cirrostratus
(a flat layer of high wispy cloud).

There are also hundreds of sub-species of clouds – some very rare, which we'll come to a little later. In the meantime, see if you can spot any of the 10 main clouds shown here outside!

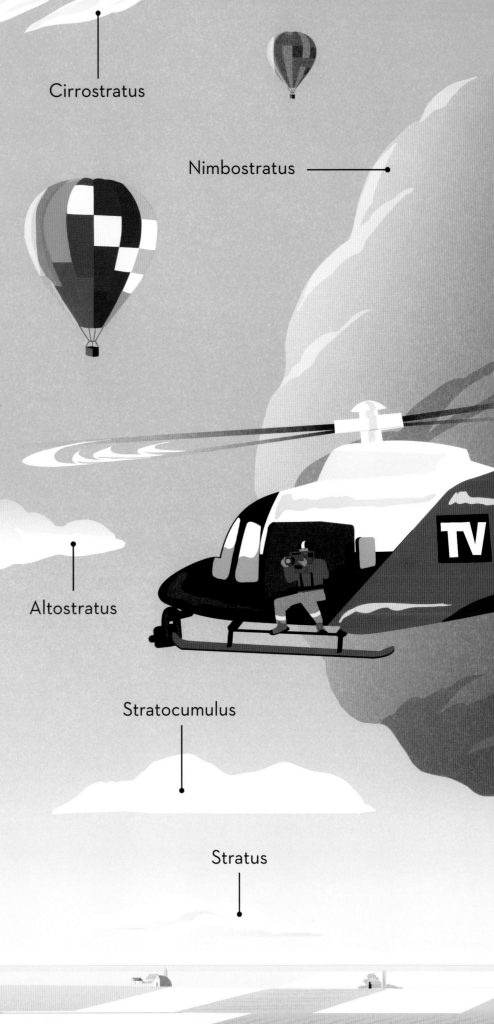

Cirrostratus

Nimbostratus

Altostratus

Stratocumulus

Stratus

WAVE CLOUDS

A HIKE OVER THE MOUNTAINS

Wind travelling around our planet has to navigate the changing landscape. Smooth waters allow it to cruise along effortlessly, but over land the friction from rough terrain forces it to slow down, change direction, rise, swirl, bounce and tumble. Mountains especially are one of the greatest influences on wind.

A Bump in the Road

When wind travels between two mountains, the squeeze makes it **accelerate**. If the mountains are too tall, the wind will even change direction and blow around them instead. But when the wind climbs a mountain, it moves up and down on the other side, bouncing like a car after a bump in the road. When this happens, 'wave clouds' can form – giving a distinct pattern across the sky.

Keeping Level

But what makes this pattern happen? In the atmosphere, there is a particular height at which clouds start to form, called the **condensation level**. This is where it's cold enough for rising moisture to condense into water droplets. However, when air sinks below this level, the moisture **evaporates** and the cloud breaks up and fades – so when the air bounces above and below this condensation level, the clouds keep forming and breaking in waves. Because the up-and-down motion of the air bouncing over a mountain often keeps going for a long way afterwards, these wave clouds can stretch for tens of kilometres beyond the mountain itself.

Brace for Turbulence

Wave clouds are relatively common and are a useful warning sign for pilots about the potential for **turbulence** – where the bouncing air could give a plane or glider a bumpy ride. It's another example of how although we can't see the wind, we know what it is doing because of the effect it is having on something else.

27

SUNRISE & SUNSET

NATURE'S FIERY ARTWORK

Colourful and mesmerising, sunrises and sunsets can offer some of the most beautiful natural displays in the skies above: fleeting, magical moments that captivate our attention – knowing that they won't last for long.

Watching the Waves

The colours that we see vary because the white light from the sun is actually made up of a spectrum of different colours: red, orange, yellow, green, blue, indigo and violet. Each of these colours has something called a **wavelength**. The wavelength determines how our eyes see it when it moves through and hits different things – such as molecules in the air, the atmosphere or surfaces around us. Red, orange and yellow light have longer wavelengths, whereas green, blue, indigo and violet light have shorter wavelengths.

As white sunlight passes through Earth's atmosphere, it collides with tiny **gas molecules** (mostly nitrogen and oxygen) in the air. These molecules are so small that they scatter the blue light towards us more than other colours. This means that we see this colour the most, making the sky look blue.

All in the Angle

During sunrise and sunset the sunlight reaches our planet at a lower angle. This leads to sunlight travelling through more of the atmosphere, hitting extra molecules along the way. Therefore, the blue light is scattered so much that less of it reaches our eyes. This allows a greater amount of the longer wavelength red, yellow and orange light to be seen, giving sunrises and sunsets their fiery appearance.

Did you know that it's not just tiny molecules of gas that can cause colourful sunrises and sunsets? Dust particles in the air from sandstorms or volcanic eruptions can also scatter light, contributing to an extravagant show of colours.

The Cloud Filter

Clouds play a part too. They can bounce light around and scatter it even more – enhancing the range of colours seen across the sky. This effect is most striking when there are different types of clouds at different heights. High, thin, wispy clouds glow brightly, as light shines through. However, thicker clouds reflect light. This creates the dramatic contrast where one side of a cloud is brilliantly lit and the other shadowed side is darker. I love it when this happens. It's when I've taken some of my most stunning pictures of the sky.

NOCTILUCENT

NACREOUS

If you're lucky enough to come across these rare clouds, make sure you get a picture – and send it to a weather presenter!

RAREST CLOUDS

Some clouds are so rare that many of us will only see them in pictures. These elusive clouds only appear briefly when conditions are perfect. However, when they do form, they stand out from the rest because of their unique appearance.

Nacreous and noctilucent clouds are rare because they form in extremely dry parts of the atmosphere: the stratosphere and mesosphere. The lack of moisture means that there aren't as many ice crystals to form clouds as lower down in the troposphere.

NACREOUS CLOUDS FORM VERY HIGH, AROUND 20-30 KILOMETRES UP, IN THE STRATOSPHERE.

They are extremely distinctive because they glow brightly with a variety of colours. Known as **iridescence**, this effect is similar to the shimmering colours that you can see on a soapy bubble. The word 'nacreous' even originates from the French word 'nacre', which is used to describe the pearly colours found on shellfish.

Nacreous clouds need a very low temperature (below minus 78 degrees Celsius) to form, so are usually only found in polar regions in winter. They are also lit up by sunlight from below the **horizon**, restricting their **visibility** to a window before dawn and after dusk. You're very lucky if you've managed to spot nacreous clouds. I'm always staring up at the sky, yet I've only ever seen them in a picture!

NOCTILUCENT CLOUDS ARE THE HIGHEST, FORMING AROUND 80 KILOMETRES UP IN THE MESOSPHERE. THIS PART OF THE ATMOSPHERE IS EVEN HIGHER AND DRIER THAN THE STRATOSPHERE.

Noctilucent clouds resemble thin, wispy cirrus clouds and have a bluish or silvery colour, sometimes with a tinge of red or orange. They're visible from the mid- to polar latitudes in both the northern and southern hemispheres.

To see noctilucent clouds when they form, you normally need a clear summer evening during twilight. This is the short time after sunset when it's not completely dark yet, and the atmosphere and clouds are still lit up by scattered sunlight – even though the sun has set. The word 'noctilucent' comes from the Latin words 'nox' (night) and 'lucentem' (to shine). Put them together, and you get night-shining clouds.

IN AWE OF THE AURORA

BOREALIS & AUSTRALIS

Welcome to the aurora borealis – a peaceful yet visually dramatic show of nature. A magical feeling, as you're greeted by a spectrum of colours that snake, streak and dance against the deep darkness of space. I've only ever seen the aurora borealis once, but it was an awesome experience that I'll never forget!

A Wandering Firefox

Hundreds of years ago, people used to think that the **aurora borealis** represented a variety of things from spirits to dragons, or even a sign that something bad was going to happen. The Finnish word for the northern lights is 'revontulet', which translates literally to 'fox fires' – named after the mythical firefox that wandered the north of the country.

A Spectacular Light Show

The trigger for these spectacular light shows originates from the sun, 150 million kilometres away, when it sends huge amounts of charged particles hurtling towards our planet – known as the **solar wind**. These particles are full of energy and travel at speeds of 1.6 million kilometres per hour.

Upon reaching Earth, the charged particles encounter our **magnetic field**, an invisible force field that protects our atmosphere from space radiation. The field guides these charged particles towards the polar regions – attracted by the pull of the magnetic north and south poles. It's here that they collide with gas molecules, and then the fun begins!

The collision of charged particles with gas molecules releases energy as light. This light ripples across the **thermosphere** and creates the aurora borealis, producing different colours depending on the altitude and the gas molecules it hits. Oxygen molecules produce green light when hit around 95 kilometres high, but above 160 kilometres they produce red light. Nitrogen produces blue light at 95 kilometres high, but pink light when hit higher. In some cases, you can see yellow and orange light too.

The aurora borealis happens in the northern hemisphere, mostly around the polar regions. However, sometimes the display is so energetic that it can be seen in mid-latitudes as well. The same thing happens in the southern hemisphere, except that it has a different name: the **aurora australis** or southern lights.

Finding the Light

Powerful satellites and telescopes can monitor the sun's activity to make space weather **predictions** on the timing and strength of the solar wind reaching Earth. Beyond predicting the northern or southern lights, it's also useful for detecting **solar flares** and **geomagnetic storms**. These are stronger bursts of energy from the sun which can damage satellites and **navigation systems**, as well as disrupt radio signals and even the power grids that supply our electricity. Certainly something to watch for!

33

THUNDERSTORMS

DARK TOWERS IN THE SKY

Shaken by deep rumbles of thunder. Startled by bright flashes of lightning.
Soaked by torrential downpours of rain. Buffeted by strong, gusty winds.
Thunderstorms produce some of the most dramatic weather on Earth.
Their towering clouds are so dark that they almost turn day into night,
and if they become severe they can unleash their most dangerous
threat: tornadoes. But first, how do thunderstorms begin?

The First Signs of Change

Thunderstorms are actually the culmination of a build-up of heat and humidity. Early in the morning, the skies may be clear with plenty of sunshine and little wind. You might think a sunny, settled day lies ahead – until the first signs of a change.

Clouds Bubbling Up

By mid-morning, as the sun climbs higher, it starts to heat the ground. The temperature increases and air starts rising. As it reaches the condensation level and cools, the moisture condenses into water droplets and fluffy white cumulus clouds appear. Despite their appearance, an average cumulus cloud weighs around 400 tonnes – the same as 100 elephants!

A Darkening Sky

As it approaches early afternoon, the sun is overhead, sending the temperature soaring. Air rises faster and reaches higher, allowing the clouds to grow bigger, taller and darker. Innocent cumulus clouds turn into threatening cumulus congestus clouds – bringing the chance of showers.

Lightning Strikes

Mid-afternoon sees the heat build to a peak. Air is rising so high (taking lots of moisture with it) that cumulonimbus clouds form, generating torrential rain and lightning. As lightning travels through the air, it rapidly heats it up. The air expands very quickly with a loud crack – and that's what we hear as thunder.

STIRRING UP TROUBLE

Sometimes thunderstorms become so severe that the air starts to viciously rotate, leading to tornadoes. Tornadoes produce the fastest winds on Earth, reaching speeds up to 500 kilometres per hour. They can be kilometres wide and travel along the ground for up to a few hundred kilometres, carving a path of destruction. Houses are flattened, cars thrown around like toys and the ground itself torn up.

Another hazard from thunderstorms is hail – balls of solid ice. Typically, they're smaller than a pea, but they can be as big as golf balls or baseballs. These can cause a lot of damage: they can smash windows, dent cars and make holes in roofs, and completely flatten crop fields.

*Funnel cloud or **tornado**? Does the rapidly rotating column of air reach the ground? If it does, then it's a tornado. If it doesn't, then it's a funnel cloud.*

TORNADO OUTBREAK

OKLAHOMA AND KANSAS

In less than 21 hours, a total of 74 tornadoes touched down across the US states of Oklahoma and Kansas on 3 May 1999. The rapidly rotating twisters carved paths of total destruction, leaving 46 people dead and around 800 injured. More than 8,000 homes were damaged or destroyed, with property damage estimated at 1.5 billion US dollars.

This large and deadly outbreak took place in a part of the United States known as Tornado Alley, in the middle of the country. It has this nickname precisely because of its high frequency of tornadoes, and it's perfectly positioned for a variety of air masses to collide. The temperature and humidity contrasts of those air masses are a recipe for severe thunderstorms – some of which spawn tornadoes.

A Trail of Destruction

One of the tornadoes reached the maximum F5 intensity on the Fujita Tornado scale, with wind speeds of around 486 kilometres per hour. Tornadoes of this intensity produce the most powerful winds on Earth – enough to rip houses from their foundations. This tornado carved up the ground for around 60 kilometres, along a path from Chickasha through the south Oklahoma City suburbs of Bridge Creek, Newcastle, Moore, Midwest City and Del City.

WEAK TORNADO

STRONG TORNADO

DEVASTATING TORNADO

Emergency Alert

People who live in tornado-prone areas are very safety-aware, always knowing what to do should one head their way. In the United States, as well as warnings in weather forecasts, there are emergency alert systems to tell people that a tornado may be approaching. This saves many lives each year.

Once a tornado risk has been detected, people shelter in basements or small rooms – staying away from windows that could throw broken glass through the air. After finding the safest place possible, they protect themselves from any chance of flying debris and wait for the storm to pass.

TORNADO RECORDS

Longest lasting single tornado: 3.5 hours duration, stretching 352.4 kilometres from Ellington, Missouri, to Princeton, Indiana, on 18 March 1925.

Widest tornado (maximum diameter): 4,184 metres, recorded in El Reno, Oklahoma, on 31 May 2011.

Highest recorded tornado wind speed: 486 kilometres per hour in Bridge Creek, Oklahoma, on 3 May 1999.

Largest tornado outbreak: 207 tornadoes in 24 hours in the southeastern US on 27 April 2011.

LIGHTNING

A SHOCK TO THE SYSTEM

Sparks of electrically charged bright light that flicker across the sky – some striking the ground, some stretching towards other clouds. Lightning is both fascinating and scary at the same time. Whilst it has the ability to momentarily turn night into day, it can be deadly and damaging if it hits an object on the ground.

Lightning is caused by the collision of ice crystals and hail within a cumulonimbus cloud in a thunderstorm, which gives the bottom of the cloud a negative charge. The ground and the top of the cloud end up with a positive charge.

Eventually, the negative and positive charges build up so much that lightning – a huge spark of electricity – is released. The lightning hits either the ground, tall objects or other clouds.

Thundering Past

As lightning travels through the air, it rapidly heats it up. This causes the air to expand and a shockwave to be created, which is what we hear as thunder.

First Light

Light travels much, much faster than sound. Therefore, it takes longer for the sound of thunder to reach us, compared to the flash of light. The further you are from the lightning flash, the longer it takes to hear the thunder.

Gone in a Flash

A lightning strike travels at a speed of around 435,000 kilometres per hour. At this speed, it would take just 46 seconds to travel from London to New York. London to Sydney would take only two minutes and 21 seconds.

Hot Shot

The electrical discharge from lightning is so intense that it can reach a temperature of 30,000 degrees Celsius, which is five times hotter than the surface of the sun!

Striking the Hour

On average, there are almost four million lightning strikes on Earth each day. That's around 158,400 every hour, or 44 strikes per second!

Blasting the Tree

When lightning strikes a tree, it is often destroyed. Lightning instantly and intensely heats up water and sap inside the tree, causing it to rapidly expand. This often causes the bark of the tree to be blown away and the wood inside to split.

BOLT-CANO

Volcanoes cause lightning as well. When they erupt, they throw up huge clouds of rock and ash into the sky. Differences in electrical charge happen in the same way as with ice in a cloud – eventually leading to lightning strikes.

PRECIPITATION

WATERING OUR PLANET

Did you know, it's sometimes possible to smell when it's raining? Yes, that's right! When rain falls on sandy or clay soils after a dry spell, the raindrops hitting the ground help to release an earthy scent into the air. It's called petrichor.

Precipitation is any form of water that falls from the sky – be it rain, snow, hail or a mixture of them all. It's nature's clever way of moving large amounts of water from one location to another. Moisture in the air rises and condenses to form clouds in one place, before they drift away and deposit it elsewhere. How this water is distributed determines the landscape and climate of an area – such as where luscious, moisture-rich rainforests or barren, parched deserts are found.

The type and amount of **precipitation** that falls in any one place can have a big effect on our daily lives. Heavy snow can look pretty and mean a day off school, but it can cause lots of disruption to travel and infrastructure – such as electricity supplies. Hail may be interesting to watch, but if the hailstones are big enough, they can damage roofs or smash windows. Impacts from precipitation tend to be greatest when there's either too much or not enough of it.

Rain

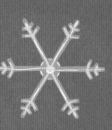

Despite rain being the most familiar type of precipitation, there are lots of interesting facts about it that people don't know!

Raindrops vary a lot in size. The smallest ones, like a fine spray, have a diameter as little as 0.2 millimetres. The biggest ones, found in tropical regions, have a diameter around 5 millimetres. It's not just their size that varies, but also their speed. Raindrops typically fall at 11-32 kilometres per hour. However, the smallest, finest drops fall more slowly, at around 5 kilometres per hour.

Snow

Snow is easily the prettiest precipitation. Snowflakes of varying shapes and sizes gently descend at around 1-6 kilometres per hour, turning the landscape into a winter wonderland. Due to the way in which ice crystals create snowflakes, they are all hexagonal – with six sides or points – but are also unique. Every flake has its own pattern. Snow is often described in two ways – either wet and sticky, which is good for making snowballs, or dry and powdery, which is good for skiing.

Hail

Hail can have quite a journey before it hits the ground. Inside a thunderstorm, hailstones are carried up and down by air currents, accumulating another layer of ice each time. This means that if you slice a hailstone in half, it resembles an onion with its individual layers.

Small hailstones fall at 15-40 kilometres per hour. But the biggest ones can fall as fast as 115 kilometres per hour!

In the most severe thunderstorms, hailstones as big as golf balls and baseballs have been recorded, which can cause a lot of damage when they reach the ground!

RAINBOWS

HOW TO SPLIT A SUNBEAM

We all know what rainbows look like, but we don't see them often, and when we do they don't appear for long. They're what we call an 'optical phenomenon': a special interaction of light and matter. So what is the recipe for a rainbow?

Light Work

White light from the sun is actually made up of a spectrum of different colours: red, orange, yellow, green, blue, indigo and violet. Each of these colours of light has a specific 'wavelength', which determines how we see them as they move through different things – such as molecules in the air, Earth's atmosphere or, in the case of rainbows, water droplets.

SHORTER ← WAVELENGTH → LONGER

Two For One

If you're really lucky, you might even see a double rainbow. These happen when light reflects twice in each water droplet, instead of once. This causes a second bigger and fainter rainbow, which arcs above the main rainbow, and has the colours in reverse order.

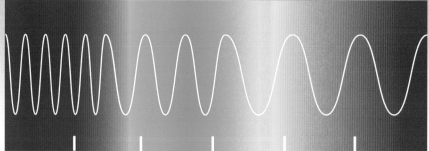

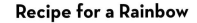

Recipe for a Rainbow

1. The sun needs to be behind you.

2. The water droplets need to be in front of you.

3. The sun needs to be low in the sky – less than 42 degrees above the horizon.

4. The best weather is a mixture of sunshine and showers to give sunlight and water droplets.

Refraction in Action

When white light hits water droplets, it slows down and changes direction – this is known as **refraction**. Refraction splits the white light into different colours. These different colours of light then hit the back of water droplets and are reflected towards our eyes – making us see a rainbow.

Upside-down Rainbows

Known as 'circumzenithal arcs', upside-down rainbows form in a similar way to rainbows – except that they don't touch the horizon and usually appear in cirrus clouds. They occur when sunlight hits ice crystals (rather than water droplets) in high clouds. The sun needs to be low in the sky – below 32 degrees above the horizon – for circumzenithal arcs to happen. This means that they tend to be seen most often during mid-morning or mid-afternoon, so keep an eye out!

MIST & FOG

WHERE THINGS GET HAZY

When grey vapour hangs in the air, the world can seem quite spooky. But fog and mist are actually just clouds that form close to or on the ground. When you're out on a foggy day, you are essentially inside a cloud.

Sometimes, fog only stretches a few hundred metres up from the ground. This means that if you go to a high floor in a tall skyscraper, it can be sunny outside and you can see the layer of fog down below.

Solving the Mist-ery

Fog and mist are formed in a very similar way to clouds. When the right conditions are present and the air cools, moisture in the air condenses into water droplets. You can see this at home when steam hits a cold mirror: the moisture condenses into drops of water.

So, if fog and mist form in the same way, what's the difference? The answer lies in how far you can see. If the visibility is 1 kilometre or more, it is mist. If the visibility is less than 1 kilometre, then it is fog. When the visibility falls below 200 metres – about the length of a 10-carriage train – it is often described as thick or dense fog. This can cause real problems for transport.

Radiation Fog

Radiation fog is most common in autumn and winter – especially in **temperate** climates. When skies are clear at night, heat in the ground escapes (radiates) so temperatures fall. If the air gets cold and calm enough, moisture in the air condenses into fog. Radiation fog can take a long time to clear in winter – sometimes lingering all day.

Advection Fog

Advection fog is most common in spring and summer, when a warm, moist air mass moves over cold ground or water. This cools the air mass, condensing its moisture into fog. The coasts of California, in the United States, are famous for this: if you've seen pictures of the Golden Gate Bridge in San Francisco shrouded in fog, that's advection fog!

Freezing Fog

In autumn and winter, when the temperature hits zero degrees Celsius or lower, freezing fog can form. The water droplets in the air freeze against surfaces, creating a coating of feathery crystals. If this happens on the ground, pavements and roads can become very slippery and dangerous.

★ FINAL ★

CHICAGO
EXPRESS

AFTERNOON
EDITION

VOL. 21. NO. 311   JANUARY 27, 1967   120 PAGES, 10 CENTS

WINTER'S WRATH

THUNDERSNOW, BLIZZARDS AND ICE STORMS

Snow is often pretty and tranquil: a winter wonderland with a crisp blanket of white. However, winter can have an awful temper too, and unleash ferocious hazards on millions of people.

THUNDERSNOW

Thundersnow is exactly what it sounds like: thunder and snow together, and just like a rainy thunderstorm produces lots of rain, thundersnow can produce lots of snow very quickly. Most people are unlikely to experience it, because it usually only happens when cold Arctic air moves over warmer water, along coastlines or over large lakes. Around the Great Lakes of North America, thundersnow is quite common in winter, where as much as 10 centimetres of snow can fall in an hour.

The warmer water heats the cold air, sending it rising into the sky to form dark cumulonimbus clouds. As ice crystals in the cloud collide, an electrical charge builds up, which discharges out as lightning. Down on the ground, you would see a bright flash through the falling snow – even brighter than in a rainstorm as the light reflects off all the snowflakes – followed by a deep but dampened rumble of thunder.

BLIZZARDS

Blizzards are when strong winds meet heavy snow, and they're one of the most dangerous winter hazards: the snow can cover everything in minutes, and a howling wind whips up snowflakes into a blinding wall. In a severe blizzard, visibility can be near zero, making travel impossible. Anyone outside risks becoming disorientated and lost.

In January 1967, a major blizzard hit Chicago in the United States. Snow fell for 29 hours, with 58 centimetres on the ground. The wind reached 85 kilometres per hour: so strong that it blew the snow into drifts as tall as doors, left 20,000 cars and 1,100 buses stranded – and shut down the airport. Hospitals had to use helicopters just to get medical supplies.

Ice Storms

It's hard to imagine everything covered in a slippery layer of ice – not just roads and pavements, but trees, houses and power lines. That's the effect of freezing rain in an ice storm. A layer of warm air gets sandwiched between two layers of cold air, causing snow to melt into rain as it falls through the middle. The rain then freezes again just as it hits everything in the cold layer at ground level. It's an instant ice-rink! Ice building up can become so heavy that trees and power lines bend or even snap in half. This is why ice storms can leave areas without electricity for weeks even after the storm has passed and the ice has melted.

WIND CHILL

AN INVISIBLE THREAT

On a cold, calm winter's day, you can feel fine outdoors if you're all wrapped up – until the wind starts blowing. This effect is called wind chill, and the faster the wind blows the colder it will feel, despite the temperature staying the same.

Sending Shivers

Wind chill is a danger we can never see. If you're caught in a cold wind, it can feel like it's blowing straight through you. First as just a chill, then a numbness in your fingers and face, and if you don't find shelter, it can even prove fatal. It feels much colder when the wind is stronger because it blows away the thin layer of warm air we usually have around our bodies.

The Speed of Cold

Wind chill is all about how fast the wind is blowing versus the actual temperature of the air: the faster the wind, the colder it feels. In weather forecasts, you'll often hear this result described as the 'wind chill' or 'feels like' temperature.

Biting Chill

If skin is exposed to extreme cold, it can lead to **frostbite** – where body tissue starts to freeze. In mild cases, frozen skin and blisters can form. In severe cases gangrene may occur, where a loss of blood supply causes body tissue to die and skin may turn dark blue or black. In cases of both frostbite and hypothermia, it is important to seek help quickly, in order to stop them from getting any worse.

This is what happened to Robert Falcon Scott – or 'Scott of the **Antarctic**' – and his team, when they raced to become the first people to reach the South Pole in the early 1910s. They managed to get to the Pole, but only to find that a team from Norway had beaten them there. Sadly, Scott's team faced extreme cold on their way back, and died from exposure before they could make it home.

Catch Your Death

In the autumn and winter months, wind chill can become life-threatening. One danger is **hypothermia**, which is when the core temperature of our bodies falls from the usual temperature of 37 degrees Celsius to below 35 degrees Celsius. Signs of hypothermia include uncontrollable shivering, memory loss, disorientation, slurred speech, drowsiness, and exhaustion.

Suit Up

When you have to go outside in the cold, the best thing you can do to protect against wind chill is to dress in many layers of clothing. This traps warm air between each layer, insulating your body. Also, don't forget a hat, gloves, scarf and something waterproof!

HUMIDITY

A LOAD OF HOT AIR

On some hot days, the air feels heavier than on others, but why? What creates the muggy feeling that weighs us down? It's called humidity.

Soaking Up the Sun

Humidity is a measure of the amount of moisture in the air: the higher the moisture level, the more humid it feels. The most significant sources of water for a humid air mass come from the tropical oceans and rainforests, where warmth and sunshine evaporate water from the oceans and increase how much water is released into the air by the leaves of plants.

Blowing Hot Air

Weather systems and wind can carry humid air from the tropics elsewhere – such as northern Europe and the United States. This is especially dangerous for people and animals, as they aren't used to such conditions. It's worse in summer: water evaporates faster as temperatures rise, which is why warmer air can contain more moisture than cooler air. This makes high humidity in hot, sunny seasons feel worse than at other times.

*During hot and humid weather, it's important to stay cool and drink enough water to avoid **dehydration** and overheating. Dehydration is when your body loses more water than it takes in. It can make us thirsty, dizzy or tired, as well as making our mouth, lips and eyes dry.*

Signs of Heat Exhaustion

If you overheat, heat stress can stop your body from being able to control its own temperature. The first sign of danger is when your body temperature reaches 38 degrees Celsius or above and heat exhaustion happens. Signs of heat exhaustion include:

☀ *Not sweating when too hot or excessive sweating.*

☀ *Being very thirsty and having pale skin.*

☀ *Cramps in your arms, legs and stomach.*

☀ *Having dizziness, confusion or a headache.*

☀ *Being short of breath or having a fast heartbeat.*

☀ *Losing consciousness.*

If someone with heat exhaustion doesn't cool down, it can turn into heatstroke, which is very serious and needs urgent treatment.

Don't Sweat It

Our bodies keep cool by sweating. When we sweat, heat evaporates the moisture on our skin into the air. This has a cooling effect and stops us from overheating. In order for the moisture to evaporate, the air surrounding us must be able to hold it. However, when the humidity is high, the air is like a **saturated** sponge – so full of moisture already that our sweat struggles to evaporate. This reduces the cooling effect, making us feel hot, humid and uncomfortable.

HEAT WAVES

SCORCHING THE EARTH

Heat waves can be so intense that it makes being outside in the afternoon dangerous. They are most common in summer, and typically last for at least three days. However, sometimes they can endure for weeks or even a month, with each day as hot or hotter than the last. So where do they come from, and why do they last so long?

Making Waves

The main ingredient for a heat wave is a big area of high pressure that sits over the same region for at least a few days. High pressure in summer tends to bring cloudless skies, allowing the sun to shine down all day long. This heats the ground, which heats the air above, sending temperatures soaring.

At the equator, the number of daylight hours each day is similar all year round, but further away from the equator, summer days are longer than summer nights. As a result, the sun's energy hits the ground for longer periods, building up the heat day after day.

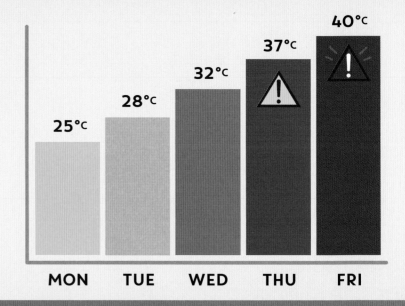

| 25°C | 28°C | 32°C | 37°C | 40°C |
| MON | TUE | WED | THU | FRI |

European Heat Wave of 2019

In summer 2019, Europe had record-breaking heat waves. Hot air from North Africa and an area of high pressure gave day after day of sunshine and heat. The nights became so warm and sweaty that millions of people struggled to sleep.

The heat broke temperature records in France, Germany, Luxembourg, the Netherlands, Belgium and the United Kingdom. For some countries, it was the first time a temperature above 40 degrees Celsius had been reached.

Thousands of people and animals died due to the heat wave. The intense heat caused major disruption to transport and electricity networks, and power stations that rely on water for cooling had to shut down, due to water shortages.

Cool Constructions

Throughout history, there have been some clever ways that people have kept buildings cool. For centuries, the walls and roofs of buildings in southern Europe and North Africa have been painted white. This isn't for decoration: light-coloured surfaces reflect the sun's rays back into the sky, so less energy is absorbed by the building, leaving it cooler inside.

Another method used in ancient palaces in India was 'evaporative cooling'. Water channels flowed through the rooms, and the heat in the air evaporated it. This lowered the temperature inside to make the rooms cooler and offer some respite through the day.

DUST STORMS

SWEEPING UP THE DESERT

Also known as sandstorms, dust storms happen when weather fronts or thunderstorms, and their strong winds, lift huge areas of dust and sand. They usually stretch 1–2 kilometres high, and move along at 40–80 kilometres per hour, so it can look like a wall barrelling towards you, consuming everything.

Sydney's Red Dawn

In September 2009, a huge dust storm swept across the eastern side of Australia. It reached 2.5 kilometres into the sky and was approximately 3,000 kilometres long. That's huge. It's as far as the distance between Spain and Iceland.

On 23 September, the dust storm arrived in Sydney. People woke up to a thick red haze blanketing the city. It quickly became known as 'Red Dawn'. The dust was so thick that at the height of the storm the visibility dropped to just 400 metres. It was estimated that the storm carried around 15 million tonnes of dust at its peak. That's the same weight as 34,000 jumbo jets.

Particularly Dangerous

As well as reducing visibility, dust storms can impact parts of the environment, along with the health of people and animals.

Some dust particles in the air are very small and can seriously affect people who suffer with breathing-related illnesses, such as asthma.

Effects on the environment stretch to both land and sea. On land, dust storms can strip away soil and its nutrients, leaving poor growing conditions for farmers' crops. In the sea, dust can increase silt in the water, which blocks the sunlight that coral reefs need to survive, as well as creating toxic algae.

Up, Up and Away!

Even when a dust storm ends, tiny particles can be whipped up so high into the sky that they are carried by the wind far away from where they came. Dust from the Sahara Desert in Africa can travel thousands of miles across the Atlantic Ocean and reach as far as the Amazon rainforest, where it can be a good thing, fertilising soil for plants to grow.

Dust can also help to suppress hurricane activity over the Atlantic Ocean. Tropical disturbances, that are the beginnings of hurricanes, need moist, humid air to grow. Very dry, dusty air hinders their growth.

Sometimes, dust gets carried into rain clouds and falls in rain, and when it evaporates, it leaves a thin coating of dust. So next time you notice your car coated in dust after rain, think how far it must have travelled to get there!

DUST DEVILS

RISING FROM THE EARTH

A swirl of sand and dust appears out of nowhere, but there's not a cloud in sight, so what is it? It's a dust devil, which is a small area of rapidly rotating wind, made visible by the dust and dirt that it picks up from the ground. They can be anywhere from one to 100 metres wide, and taller than skyscrapers. Many of them have spinning winds of around 70 kilometres per hour. Usually they cause few problems, but sometimes they reach wind speeds as high as 120 kilometres per hour – as strong as a weak tornado – causing lots of damage.

Dust Devil Hotspots

Dry, sunny and hot climates usually make it hard for things to grow, so it can cause land to be barren. Land that doesn't have many plants often becomes dusty or sandy: the perfect conditions for dust devils!

Sandy surfaces heat up more quickly and intensely than grass and damp soil, which makes them favourable places for dust devil formation. The same is true for tarmac, so they can even happen in towns and cities!

Dust devils are sometimes confused with tornadoes, but there's a big difference: tornadoes form from clouds and reach down to the ground. Dust devils reach up from the ground to the sky. Easy-peasy!

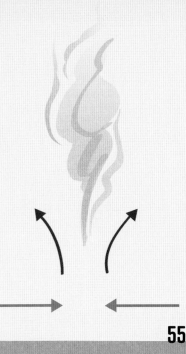

How Do They Form?

On hot, sunny days when there is no wind, the ground is intensely heated by the sun. However, some parts may be in direct sunlight and other parts in the shade, so certain areas end up hotter than others.

Hotter air over an intensely heated area of ground rises into the sky easier than nearby cooler air. When this happens, a mini area of low pressure forms. A dust devil occurs when this rising column of air starts to rotate, reaching higher into the sky – becoming more visible as it whips up dust, sand and soil from the ground.

A dust devil only lasts a few minutes. Eventually, as the hot air rises, cooler air gets drawn in at the bottom to replace it, cutting off the supply of hot air. It breaks up in seconds.

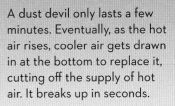

WEATHER BOMBS

WHEN PRESSURE DIVES

A weather bomb or bomb cyclone, also known as 'explosive cyclogenesis',
is when a low-pressure system develops more intensely and more quickly than usual.
They typically happen at mid-latitudes (e.g. around Europe and North America) over the
world's oceans in autumn and winter, producing some of the stormiest weather of the year.
From space, a weather bomb might look like a beautiful, swirling pattern of cloud,
but when it moves over land it leaves behind a trail of damage and disruption.

EXPERIENCING A WEATHER BOMB

1. It's dry with sunshine and a light wind, although some high cirrus clouds start to appear in the sky.

2. It's overcast with rain that is light at first, but becomes heavier and steadier with the wind starting to pick up.

3. The rain turns light and patchy before stopping, but it stays quite cloudy with a brisk, steady wind.

4. Towering clouds, including cumulonimbus, bring heavy downpours – sometimes with hail and thunder. The wind turns stronger with sudden gusts as the downpours move through.

5. Skies are bright with sunshine and heavy showers; however, the winds are at their strongest, with damaging gusts that can reach 80–160 kilometres per hour.

Countdown Pressure

Weather bombs are caused by a fast-moving jet stream around 9 kilometres up in the sky. Moving at speeds of 290 kilometres per hour or more, as fast as a high-speed train, the jet stream lifts air up into the sky at a greater rate than usual. This leads to air pressure rapidly falling at the surface and a powerful area of low pressure forming.

You may have noticed on weather charts that areas of low pressure have numbers next to them. This is a measure of their central air pressure – usually in millibars (mb) or hectopascals (hPa). The lower this number, the deeper and more violent the low pressure is, and the more severe the weather (wind and rain) it produces.

In order for an area of low pressure to be a weather bomb, the central pressure needs to fall by 24 millibars or more in 24 hours. For more typical areas of low pressure, this is usually around 12 millibars or less in 24 hours.

Taking the Hit

The stormy conditions that weather bombs bring are probably the most common type of severe weather that people at mid-latitudes experience. They are particularly dangerous as they move very swiftly, meaning that the strongest winds and heaviest rain can arrive quite suddenly – taking everyone by surprise.

Although the heavy rain they produce can cause flooding, it's the wind that often causes the most damage and disruption, blowing down trees and power lines, disrupting transport and causing injuries – even deaths.

The most potent weather bombs can produce winds that are as strong as a hurricane, so they're referred to as 'hurricane-force winds'. However, weather bombs are not hurricanes: hurricanes are tropical storms driven by warm ocean water (see p.18–19), whereas weather bombs are driven by jet streams.

THE GREAT STORM OF 1987

CASE STUDY

When the people of southeast England woke up on the 16 October, 1987, they were greeted with scenes of utter chaos and destruction: the aftermath of the worst storm since 1703. It had peeled off roofs like tin can lids and littered the landscape with fallen trees.

The storm killed 18 people in the United Kingdom and four in France, where it also had a big impact. It blew down 15 million trees, including some historic specimens in London's Kew Gardens. In northern France, the storm left 1.2 million homes without electricity.

In London, gusts of 151 kilometres per hour were recorded in the early hours of the morning. But the strongest wind gust in the United Kingdom was in Shoreham-by-Sea, West Sussex, reaching 185 kilometres per hour. An even stronger gust of 216 kilometres per hour was recorded on the north coast of France, at Pointe du Roc (a cliff-top location) in Granville, Normandy.

STING JETS
WIND WITH A BOOST

Another thing that made the winds so extreme during The Great Storm of 1987 is something called a **sting jet** – a phenomenon that turbo-boosts winds to make them even more severe.

Veering from the Path

During the days before the storm, tropical maritime and polar maritime air masses collided over the Atlantic Ocean, creating a big temperature contrast. Combined with a powerful jet stream, this led to rapidly rising air – the perfect ingredients for a deep area of low pressure.

The storm also took a much more southerly track than usual, taking the severe weather across southern England, rather than a typical path towards Scotland and Northern Ireland. This meant that not only was the storm more powerful, it also hit an area that wasn't used to such strong and damaging winds.

The Hurricane That Wasn't

It's common to hear people describe The Great Storm of 1987 as a hurricane, but it wasn't – just a very deep area of low pressure. Hurricanes need a sea surface temperature above 26.5 degrees Celsius, and anyone who's dipped their toes in the sea around the United Kingdom will tell you that the water is definitely nowhere near that warm! However, the storm did locally produce hurricane-force winds.

A sting jet is a strong core of wind that forms at an altitude of 3-4 kilometres. Cold air at this height is made even colder by rain and snow evaporating as it falls through it. The air descends and accelerates, making wind gusts even stronger as it reaches the ground – sometimes more than 160 kilometres per hour.

Sting jets often create a distinctive hook pattern in the cloud, which looks like a scorpion's tail. So, if you ever see a storm on a satellite picture with these characteristics, you may have spotted a sting jet in action!

SATELLITES

KEEPING AN EYE ON THE WEATHER

Satellites are amazing pieces of equipment floating in space, constantly taking images of our planet below. They've been up there since 1960, when NASA launched the world's first successful weather satellite: TIROS-1. It gave weather forecasters their first-ever view of the clouds from space.

Light or Heat

Satellites can provide a variety of images, but two of the most common types are visible and infared – each with their own distinct characteristics.

Scanning the Sky

Different types of cloud have different temperatures. By analysing the temperatures and patterns in an image, we can tell what type of cloud has formed.

Visible satellite images are effectively like having a big camera in space and taking a snapshot of our planet below. They show what we would see with our eyes if we were to look down from space.

*Infrared images capture **infrared radiation**, which is a type of energy that we cannot see with our eyes, but we can feel as heat. This shows us the temperature of things – including clouds. On a typical black and white infrared satellite image, colder objects appear bright and white, whereas warmer objects appear dull and dark.*

Cumulonimbus clouds, which produce thunderstorms, are cold because they reach high in the sky and are made from lots of ice crystals. Therefore, they appear bright and white on infrared images.

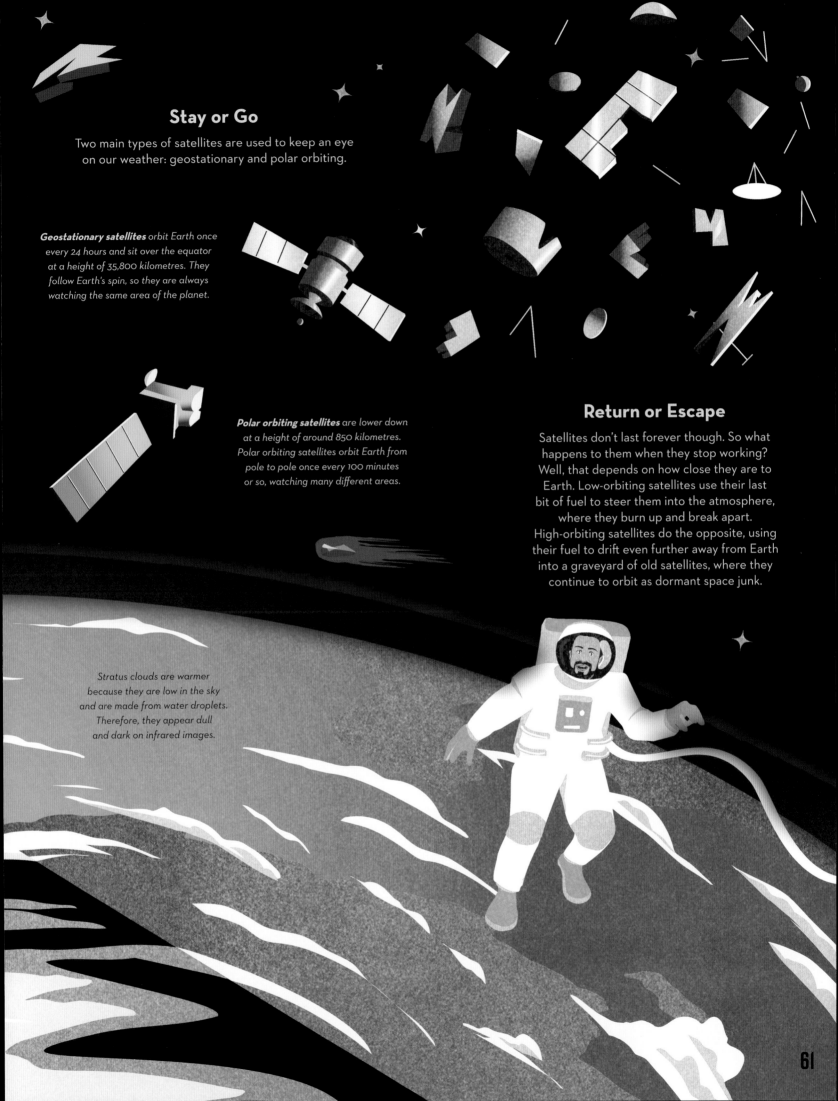

Stay or Go

Two main types of satellites are used to keep an eye on our weather: geostationary and polar orbiting.

Geostationary satellites orbit Earth once every 24 hours and sit over the equator at a height of 35,800 kilometres. They follow Earth's spin, so they are always watching the same area of the planet.

Polar orbiting satellites are lower down at a height of around 850 kilometres. Polar orbiting satellites orbit Earth from pole to pole once every 100 minutes or so, watching many different areas.

Return or Escape

Satellites don't last forever though. So what happens to them when they stop working? Well, that depends on how close they are to Earth. Low-orbiting satellites use their last bit of fuel to steer them into the atmosphere, where they burn up and break apart. High-orbiting satellites do the opposite, using their fuel to drift even further away from Earth into a graveyard of old satellites, where they continue to orbit as dormant space junk.

Stratus clouds are warmer because they are low in the sky and are made from water droplets. Therefore, they appear dull and dark on infrared images.

NUMBER-CRUNCHING DATA

HOW A SUPERCOMPUTER PREDICTS THE WEATHER

Predicting what the weather will do in the future helps us to plan our lives. Without a supercomputer, it wouldn't be possible to gather and process the vast amount of data needed to produce an accurate and reliable weather forecast.

Just How Super?

Most of us are used to computers that sit on top of our desks. **Supercomputers** fill big rooms, can be 10–20 metres in length and are more than 10,000 times faster than the average home computer. These powerful machines are capable of carrying out trillions of calculations each second. If all the humans on our planet were working together to match the calculation speed of a single supercomputer, everyone would need to do hundreds of calculations per second to keep up!

Supercomputers, owned and run by national weather services, have a massive amount of memory to store the vast volume of information they produce. But how do these impressive machines predict the weather?

For a supercomputer to make a prediction, it needs to know what the weather is currently doing. This is where weather observations play their part.

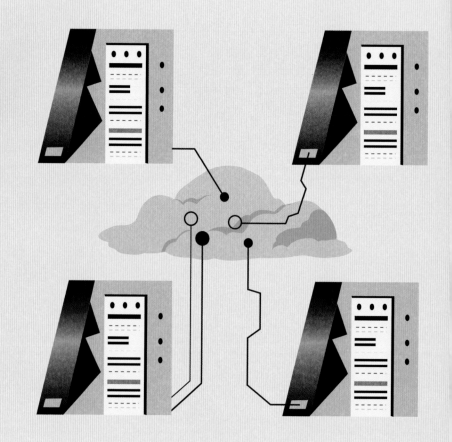

Weather Observations

The temperature of the air and ground, measured by thermometers.

Humidity, measured by a hygrometer.

Air pressure, measured by a barometer.

Cloud cover, measured by satellites.

Wind speed on the ground, measured by an anemometer.

Jet stream wind speeds, recorded by planes.

Ocean temperatures, measured by buoys.

Rainfall, measured in a container called a rain gauge.

Each day, weather observation data is fed into the supercomputer – a process called **data assimilation** – allowing it to construct a global picture of the weather.

The Old Ways

Before supercomputers were invented in the 1950s, the ability to forecast the weather relied on weather observations being taken worldwide by observatories and sent to forecasters by **telegraph** – at a speed of up to 70 words per minute. Forecasters would then plot hand-drawn weather maps of the observations and use their knowledge to figure out what would be most likely to happen next. However, the weather forecasts weren't anywhere near as accurate as they are today, and weren't able to predict as far into the future – a day ahead at most.

Making the Prediction

Once the supercomputer has a global snapshot, it runs a program known as a **numerical weather model**. This carries out billions of mathematical calculations to work out how the weather will change in the future, based on the information it had to start with.

As Earth is so big, the weather model divides the atmosphere into lots of boxes, making predictions for each one and stitching the results together. The end result is a prediction of what the weather will do in the future, typically at hourly intervals and as far ahead as 10 days.

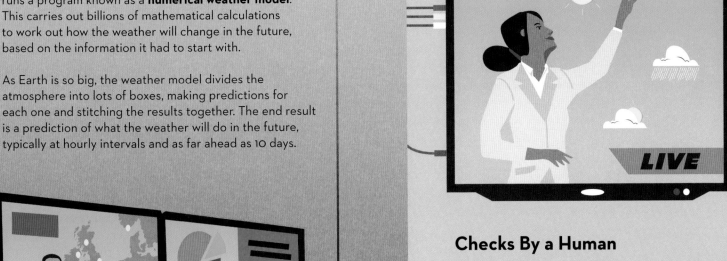

Checks By a Human

When the supercomputer has finished making its prediction, the data is checked by a human weather forecaster. This is because there are some weather patterns that weather models don't predict very well. A forecaster can tweak the prediction to make it better.

This step doesn't always happen with weather app data, so use them cautiously, as they're often automated. It means an app can sometimes show a different forecast to the one on TV.

Once the weather forecast has been checked, it's then **tailored** for use in lots of different places that need to know about different data; on TV, for emergency services, airlines, outdoor events, sailing and much more.

WEATHER CHARTS

GETTING THE BIG PICTURE

Weather charts are really handy tools for weather forecasters, as they give a quick overview of the general weather patterns over large areas for the next few days. They're useful to spot notable features such as deep areas of low pressure that create wind and rain, or big areas of high pressure that bring sunshine.

Weather charts are made using data produced by supercomputers. Most of us have seen one in a geography lesson or on TV. They can seem overwhelming and loaded with information, but after reading this page, you'll be able to read them too!

Weather Charts Guide

Weather fronts mark the boundaries between different air masses. As well as bringing a change in temperature and humidity, they tend to bring cloud and some precipitation along with them. If weather fronts are weak, they can just be bands of cloud with very little change in temperature. However, if they are strong, they can bring severe weather, heavy precipitation, strong winds and large temperature changes.

Occluded fronts are lines with both triangles and semi-circles along them – usually coloured pink or purple. Behind an occluded front is a cooler air mass and temperatures tend to fall.

For all of the weather fronts mentioned, the shapes along them point in the direction they are moving.

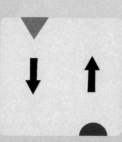

Cold fronts are lines with blue triangles along them. Behind a cold front is a colder air mass and temperatures fall.

Troughs are black lines and, unlike weather fronts, they usually only show where areas of precipitation are likely to be, rather than a change in air mass and temperature.

Warm fronts are lines with red semi-circles along them. Behind a warm front is a warmer air mass and temperatures rise.

The continuous parallel lines on weather maps are called isobars, and join up places with the same air pressure – shown by the numbers on the lines.

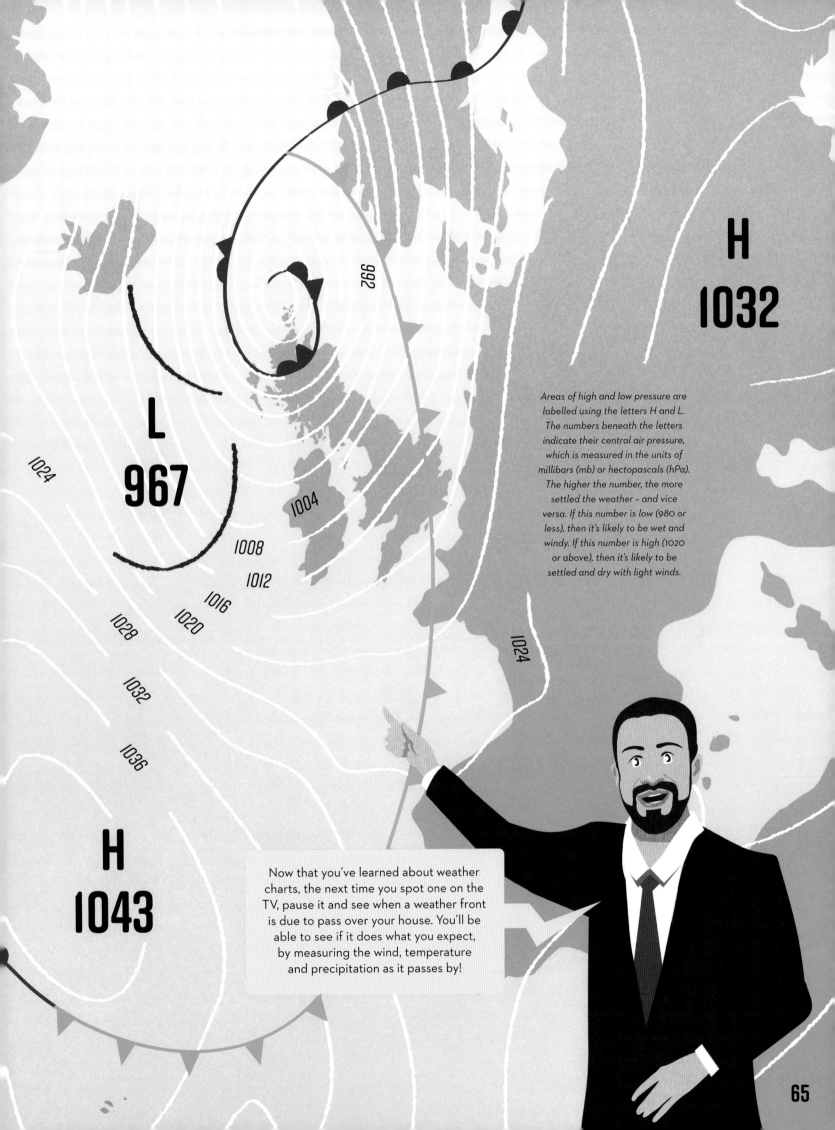

H
1032

992

L
967

1024

1004

1008

1012

1016

1020

1028

1032

1036

1024

H
1043

Areas of high and low pressure are labelled using the letters H and L. The numbers beneath the letters indicate their central air pressure, which is measured in the units of millibars (mb) or hectopascals (hPa). The higher the number, the more settled the weather – and vice versa. If this number is low (980 or less), then it's likely to be wet and windy. If this number is high (1020 or above), then it's likely to be settled and dry with light winds.

Now that you've learned about weather charts, the next time you spot one on the TV, pause it and see when a weather front is due to pass over your house. You'll be able to see if it does what you expect, by measuring the wind, temperature and precipitation as it passes by!

WEATHER FORECASTS

PREDICTING THE FUTURE

Time Is Crucial

So, with supercomputers producing the data and meteorologists using their expertise, how far ahead can the weather be predicted? Is next week's weather just as certain as tomorrow's, and when does a forecast just become a guess? The simple answer is that the further ahead into the future you look, the harder it becomes. As you've read, supercomputers and numerical weather models start with knowing what the weather is doing now, before making a prediction about the future. The difficulty is that they still only give us a good representation, rather than a perfect snapshot, of Earth's atmosphere. Weather observations are sparse in some areas, like the deserts, rainforests, or the Poles, so weather models have to fill in the gaps, which can lead to errors.

Even though these errors can start off quite small, they become bigger the further into the future of the forecast you go. The error will be used in one calculation, which will then be used in another calculation, and then another.

For example, a numerical weather model may have underestimated how cold it is somewhere and have the temperature a few degrees Celsius too high. This may not sound important, but when it uses this information to work out what type of precipitation will fall in a few days' time, it could be the difference between a snow storm instead of a rainy day during winter!

1–2 Days Ahead

Weather forecasts looking ahead for a day or two are very accurate most of the time. This is because the numerical weather models making them produce **high-resolution** data. These forecasts are able to say how the weather will change from hour to hour for a location or area.

3–10 Days Ahead

At this stage, numerical weather models used to make the forecast have a lower resolution than those for one to two days ahead. This means that they don't provide quite as much detail – due to limitations of computing power, as well as growing errors. These forecasts are fairly accurate up to five days ahead, but after that, the accuracy levels can drop significantly.

Two Weeks or More Ahead

These forecasts suffer from resolution and error problems too. However, they are useful for picking out general trends – such as whether it's likely to be warmer, or colder and wetter, or drier than average. This may not sound very useful, but it can highlight the possibility of extreme weather, like a heat wave or drought. One thing's for sure... if you ever see someone claiming to know exactly what the weather will do on a specific day weeks into the future, you'll know that it's probably a load of nonsense!

WEATHER PRESENTERS

A DAY IN THE LIFE

Weather forecasts on TV are usually only a few minutes long, so you're probably wondering what the person presenting them does for the rest of the day? Lots! So let me take you through a day in my life as a weather presenter, or broadcast meteorologist, as we are also known.

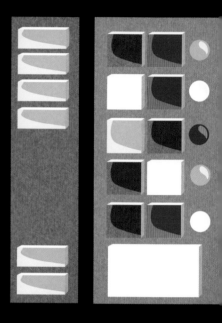

7am

I can hear the rain outside – just as I predicted yesterday. Good to see my forecast has gone to plan!

9am

The first and most important part of my day is spent on research to create a weather story to tell on the TV. This involves studying satellite images and weather observations to find out what the weather has been doing recently.

11am

The next step is predicting the future using numerical weather model information produced by supercomputers. Sometimes, it even involves having a discussion with another meteorologist.

"Next week looks very windy. Do you think there may be a weather warning issued?"

1pm

Making weather graphics is a really fun part of my day, where the information that I've researched gets turned into exciting pictures and graphics that you see on TV.

I use a special computer program that can animate a variety of weather data on a map to show how the weather changes over time.

Pictures and videos that TV viewers have taken can also be included, which is useful to show what's been happening outside. In summer, it might be people in the sunshine at the seaside, whereas in winter, it might be people playing in the snow.

Once the weather graphics are ready, it's studio time – the most exciting part of my day! The TV studio is quite a big room, with lots of lights, a camera and a special green screen.

4pm

4:01pm

The green screen is there because the computer projects the weather graphics wherever the camera sees the colour green. This does, however, mean that I can't wear green, otherwise I'll disappear into the map. Some places use a blue screen instead of a green screen, but it works in exactly the same way.

4:02pm

If there's just a green screen behind me, then how do I know where to point? Well, I see myself on a screen attached to the TV camera. Effectively, I'm watching myself present the weather as I'm actually doing it!

As well as wearing a microphone so I can be heard, I have to wear an earpiece. This allows someone to tell me in my ear how long I have left before I need to stop talking.

4:03pm

A fascinating thing that most people don't realise, is that weather presenters don't read a script on a teleprompter, as we need to see what the graphics are doing. Instead, we memorise what we have to say – all whilst appearing calm and friendly. Isn't that amazing?! We even control our own graphics with a remote-control clicker.

So, are you impressed and fancy becoming a weather presenter yourself? Maths, physics and geography are really useful subjects, along with good communication skills, of course. Being live on TV can be daunting, but as with anything, practice makes perfect!

AIR POLLUTION

A HAZARD TO HEALTH

Air pollution is a big problem around the world and the cause of around seven million deaths each year. Vehicles, homes, power generation, agriculture and industry release a mix of sulphur dioxide, nitrogen oxides and particulate matter into the air. Pollution is most common in cities, where more than 50 per cent of the world's population live.

The Weather's Influence

Even though the weather doesn't create the air pollution, it does determine how much it builds up in the air and where it goes.

If there's low pressure in the area, along with some wind, then any pollution can rise into the sky easily, disperse and move away.

However, if there's high pressure, pollution is much more likely to build up, because there's often little wind to disperse it. Also, high pressure causes air to sink, acting as a lid on the atmosphere. This traps the pollution close to the surface.

If this happens in summer, heat and sunlight can cause chemical interactions in the pollution, creating another pollutant called **ozone**. Ozone can slow plant growth, as well as make plants more at risk from disease.

LONDON'S GREAT SMOG OF 1952

The Great **Smog** of December 1952 happened at a time when coal was still being heavily used in homes for fires to keep warm, and in industries to generate energy. This produced a lot of smoke, along with other gases – all of which heavily polluted the air.

A big area of high pressure formed, which trapped the pollution near the ground, where it built up for a few days. The mix of fog and dangerously polluted air killed around 4,000 people, with many of the survivors suffering breathing problems. Travel was disrupted too, as the thick smog made it very difficult for drivers to see.

In order to stop this happening again, a clean air law was introduced to ban **emissions** of black smoke and force people to use smokeless fuels, which are less polluting.

POLLUTION FROM AFAR

Did you know that some sources of air pollution can originate from very far away? Raging wildfires – sometimes the size of cities – generate lots of smoke, as they burn through huge areas of forest and vegetation.

This smoke can get picked up by the wind and carried hundreds, or even thousands, of miles away, giving poor air quality and hazy skies wherever it ends up.

Air quality can really affect our health. When air pollution increases, adults and children with lung or heart conditions are more at risk of becoming ill and needing treatment. Breathing polluted air can irritate your airways, causing shortness of breath, coughing and wheezing. It can also trigger conditions such as asthma.

CLIMATE CHANGE

EARTH UNDER THREAT

Climate change is the biggest challenge facing life on our planet. Rising levels of greenhouse gases from human activity – especially carbon dioxide and methane – are causing Earth's atmosphere and oceans to warm up. Global temperatures have already risen by around 1 degree Celsius since 1850. This may not sound like much, but it's already having a huge impact on the weather, landscape and environment.

Warming Greenhouse Effect

The reason that these gases are described as **greenhouse gases** is that they trap heat (energy from the sun), which then warms up the atmosphere and oceans – like a garden greenhouse does to keep air warm for plants.

Earth needs a **greenhouse effect** to support life. The problem is that increases in greenhouse gases have enhanced this warming effect. This is why global temperatures have been rising so quickly since the **Industrial Revolution** – in tandem with rising carbon dioxide levels. This has allowed scientists to rule out that such a significant warming is happening naturally.

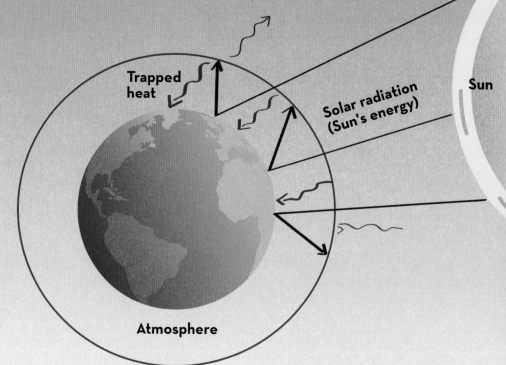

Trapped heat

Solar radiation (Sun's energy)

Sun

Atmosphere

Weather or Climate?

People often wonder what the difference is between weather and climate. Weather is the day-to-day change in patterns of temperature, rainfall and other conditions. Climate is how averages of these day-to-day weather patterns change over longer periods – decades and centuries.

No Time to Waste

Climate scientists have shown that rising global temperatures are changing weather patterns around the world, and making the weather more extreme. This is having a massive impact on people and the environment. As a result, nearly all countries have agreed to reduce greenhouse gas emissions to slow down and limit how much more our planet warms.

Everyone's Challenge

Even though the amount of warming due to climate change varies from place to place, that doesn't matter. As you've seen, the weather is **interconnected**. What begins in one place can end up thousands of miles away.

Climate change is affecting us all, right here, right now. Given the challenges we face, the need to act has never been greater. Planet Earth is our home, and it's the only home we have.

EFFECTS OF CLIMATE CHANGE

Hotter heat waves – causing increased heat stress on people and animals, and a greater wildfire risk.

Worsening droughts – risk to water supplies and food production.

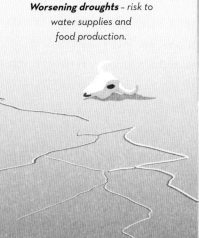

More melting of ice – rising sea levels and increased flood threat to coastal communities.

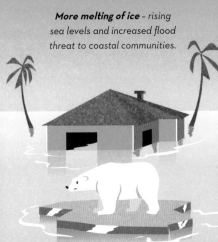

More extreme rainfall – increasing the risk of flooding.

Changes in ocean currents – changing weather patterns and the marine environment.

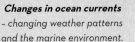

Warming oceans – damaging marine ecosystems and providing extra energy for more severe storms.

Stronger hurricanes, tropical cyclones and typhoons – increasing danger to life and property.

Warmer atmosphere – expands the **habitable** region of pests, such as mosquitoes.

A BRIGHT OUTLOOK

PROTECTING OUR PLANET

Climate change may seem like an overwhelming issue, making our weather more extreme and putting us in danger, but we are more than capable of solving it. We just need to work together and each do our bit to make changes that will help reduce the amount of greenhouse gases being released into the air. There are so many things that we can do in our everyday lives that will help.

Stop **deforestation** and plant more vegetation and trees.

Plants absorb carbon dioxide from the air and use it, alongside sunlight and water, to produce food.

Walk or cycle instead of travelling by car.

This leads to less pollution and carbon dioxide being produced.

Turn off lights and devices when they aren't being used.

This saves energy which can involve carbon dioxide being produced to generate it.

Switch to green energy, such as solar and wind power.

This is a much cleaner source of energy as no greenhouse gases are produced.

Consider eating less meat.

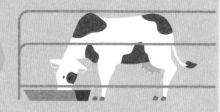

Farming animals requires the use of land, energy and fertilisers, which produce greenhouse gases. Even the burps and farts of the animals themselves produce methane!

Fly less, if possible.

The engines of planes produce a significant amount of carbon dioxide, so perhaps try to take a train instead.

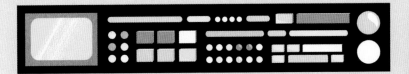

SAT_VIEW 01

Looking Ahead

Our journey through the skies has come to an end – and what a fantastic journey it has been. From raindrops to satellites, monsoons to dust devils, lightning to rainbows. Even though Earth is a huge place, its weather is so interconnected. The weather that affects you one day can affect someone else thousands of miles away a few days later. How incredible is that?!

WARNING!

INCOMING STORM

145 KPH

06:00

So, what will our relationship with the weather be like in the future? Will we all have supercomputers in our pockets that allow us to create our own personal forecasts? Will we be able to predict the weather down to the nearest minute or second? Perhaps we will even be able to reliably control certain aspects of the weather itself.

Whatever the future holds, weather will always be happening. There'll still be clouds whizzing around Earth on jet streams.

We'll still be wondering whether it'll be sunny enough to go for a walk in the park. One thing I'm absolutely sure about, is that we'll all still be talking about the weather as much as we do now!

MON	TUE	WED
10°C	12°C	11°C
THU	FRI	SAT
7°C	13°C	8°C

GLOSSARY

The words and terms listed below appear in
bold on their first use in the book.

Accelerate – to move faster.

Air mass – a body of air with a similar temperature and
humidity that covers a large area.

Air pressure – the weight of air molecules pushing down
on the ground.

Altitude – the height of an object above the ground.

Antarctic – relating to the South Pole.

Arctic – relating to the North Pole.

Atmosphere – layers of gases surrounding a planet.

Aurora australis – a natural show of bright (often green) light
high in Earth's atmosphere in the southern hemisphere.

Aurora borealis – a natural show of bright (often green) light
high in Earth's atmosphere in the northern hemisphere.

Blizzard – when strong winds and snow happen at the same time.

Catastrophic – causing sudden and major damage.

Climate change – how weather patterns change over a long
time: decades and centuries.

Climate zone – the type of weather usually experienced in
different regions of Earth.

Condensation – the change of a substance from gas to liquid.

Condensation level – the height in the sky where clouds start to form.

Continental – relating to a continent.

Coriolis effect – the deflection of wind due to Earth's spinning motion.

Data assimilation – the process of taking in information.

Deforestation – the cutting down of lots of trees.

Dehydration – when our bodies lose more water than they take in.

Doldrums – an area close to the equator with light
and unpredictable winds.

Emissions – the release of substances into the air.

Equator – an imaginary line around Earth that divides the
northern and southern hemispheres.

Equatorial – relating to the equator and the surrounding area.

Evacuate – to remove someone from danger to a safer place.

Evaporate – the change of a substance from liquid to gas.

Extreme – something very intense that has a big impact.

Fish storm – a hurricane, tropical cyclone or typhoon that
stays over the sea and never hits land.

Flash floods – flooding that happens very quickly.

Frost bite – damage to skin and body parts due to extreme cold.

Gas molecules – tiny elements that a gas is made of.

Genera – groups of things that are similar.

Geostationary satellite – a satellite that constantly
observes the same region of Earth.

Greenhouse effect – a warming of Earth's atmosphere
caused by greenhouse gases.

Greenhouse gases – gases in Earth's atmosphere that
trap heat and have a warming effect.

Gusts – sudden strong bursts of wind.

Habitable – suitable to live in.

Heat waves – prolonged periods of hot weather.

Heatstroke – illness caused by very hot weather.

High pressure – an area of sinking air that brings settled weather.

High-resolution – showing a lot of detail.

Horizon – where the ground and sky appear to meet in the
far distance.

Humid – a stuffy, sweaty feeling caused by lots of moisture in the air.

Hurricane (also typhoon and tropical cyclone) – a powerful storm
with sustained winds of 119 kph (74 mph) or more.

Hypothermia – an illness caused by very cold weather.

Industrial Revolution – the period when manufacturing, machines
and factories became widespread from the late 1700s.

Infrared radiation – a type of energy we can't see with
our eyes but can feel as heat.

Interconnected – things that are connected to each other.

Intertropical convergence zone (ITCZ) – a zone of converging
winds and low pressure in the tropics.

Iridescence – a shimmering rainbow of colours.

Jet streams – fast-moving winds high in the sky where
aeroplanes fly.

Landslide – an area of soil and/or rock that collapses.

Levees – man-made structures to prevent the overflowing of a river.

Low pressure – an area of rising air that brings unsettled weather.

Magnetic field – an invisible force field that protects Earth's atmosphere from space radiation.

Making landfall – when the centre of a hurricane, tropical cyclone or typhoon reaches land after being over the sea.

Maritime – relating to the sea.

Mesosphere – the third-lowest layer of Earth's atmosphere, around 50–80km above the surface.

Meteorologist – someone who studies and predicts the weather.

Mid-latitudes – the regions between the Tropic of Cancer and the Arctic in the northern hemisphere, and the Tropic of Capricorn and the Antarctic in the southern hemisphere.

Monsoon – a seasonal wind that brings the same type of weather for a long period.

Navigation systems – equipment that helps to guide something or someone to a place.

Northern hemisphere – the northern half of Earth between the equator and North Pole.

Numerical weather model – a computer program that uses observed weather data to make a weather forecast.

Optical phenomenon – an interaction of light and matter to create something you can see.

Orbit – to follow a curved path around a star, planet or moon.

Ozone – a polluting gas formed from sunlight, oxygen and other pollution.

Polar – relating to the North or South Pole.

Polar orbiting satellite – a satellite that observes different regions of Earth over time.

Precipitation – any form of water (solid or liquid) that falls from the sky.

Prediction – a calculation about what will happen in the future.

Refraction – the change in direction of light when it hits something, for example, water, ice or glass.

Rough terrain – land that is covered by lots of hills and mountains.

Satellite – an object orbiting Earth for communication or collecting information.

Saturated – holding as much moisture as can be absorbed.

Sea surface temperature – the temperature of the water near the surface of the sea.

Smog – fog combined with pollution.

Solar flares/geomagnetic storms – strong bursts of energy from the sun that can damage satellites and disrupt radio signals.

Solar wind – fast-moving charged particles from the sun.

Southern hemisphere – the southern half of Earth between the equator and South Pole.

Sting jet – a core of strong wind high in the sky that descends and boosts wind speed at the surface.

Storm surge – flooding caused by sea water being pushed inland by a powerful storm.

Stratosphere – the second-lowest layer of Earth's atmosphere, around 12–50km above the surface.

Subtropical – relating to the area close to the tropics.

Supercomputer – a big computer that is thousands of times more powerful than a home computer.

Tailored – to make suitable for a particular purpose.

Telegraph – a historic form of communication by sending messages over a long wire.

Temperate – a moderate climate that tends not to have extremes of hot or cold very often.

Thermosphere – the fourth-lowest layer of Earth's atmosphere, around 80–700km above the surface.

Tornado – a violent and destructive area of rapidly rotating wind.

Torrential rain – very heavy rain.

Tropic of Cancer – an imaginary line around Earth, 2,600 km north of the equator, marking the northern edge of the tropical climate zone.

Tropic of Capricorn – an imaginary line around Earth, 2,600 km south of the equator, marking the southern edge of the tropical climate zone.

Tropical – relating to the tropics.

Tropical depression – an area of cloud, rain and thunderstorms with brisk, gusty winds.

Tropical storm – an area of cloud, rain and thunderstorms with very strong, gusty winds.

Troposphere – the lowest layer of Earth's atmosphere, from the surface up to around 12km.

Turbulence – sudden and unsteady movement of air.

Vapour – tiny bits of a gas or liquid that are suspended in the air, for example, water droplets.

Visibility – how far you can see.

Waterlogged – full of water.

Wavelength – the distance over which a wave's shape repeats.

Weather bomb/explosive cyclogenesis – an area of low pressure that develops more intensely and more quickly than usual.

Weather front (cold, warm or occluded fronts) – a line marking the boundary between different air masses.

World Meteorological Organization (WMO) – a worldwide organisation that looks at weather, climate, water resources and the environment.